# *The Einstein - Schrödinger - Hawking Theory of Quantum Gravity*

FSC
www.fsc.org
MIX
Paperi vastuul -
lisista lähteistä
Paper from
responsible sources
FSC® C105338

***Dedicated***

***to***

***My wife Maria***

***and***

***Our three children.***

*Janne Pura*

# *The Einstein – Schrödinger – Hawking Theory of Quantum Gravity*

***Cover: ©DaVinci Artificial Intelligence***

***Kustantaja: BoD - Books on Demand, Helsinki, Suomi***
***Valmistaja: BoD - Books on Demand, Norderstedt, Saksa***

**ISBN 978-952-80-2381-4**

## *Foreword*

***This book is a layman´s story about quantum mechanics, gravity and cosmology based on mathematical intuition.***

***Janne Pura***

***26th September 2023 in Turku.***

***"The supreme goal of every theory is to make the irreducible basic elements as simple and as few as possible without having to surrender the adequate representation of simple datum of experience"***

***– Albert Einstein***

# *Quantum Cosmology*

***"...we may be able to understand the big bang or big crunch (in particular, whether time has beginning or an end at these events) when we have satisfactory quantum theory of gravitation"***

***– Jamal Nazrul Islam***

***Quantum cosmology is a field of theoretical physics to develop a quantum theory of the universe.***

***Quantum cosmology tries to find answers related to the first phases of the universe.***

***Einstein´s General Theory of Relativity describes the evolution of the universe but it is not quantum theory of time and space.***

***If you approach the Big Bang the General Relativity does not provide answer to such questions as gravitational singularity or the earliest events of the universe.***

*The mathematical functions in the classical theory of general relativity are continuous but quantum theory is discrete with fundamental units known as quants.*

*This is called quantization.*

*In quantum cosmology, the universe is treated as a wave function instead of classical four-dimensional spacetime.*

*Wheeler-DeWitt equation also called Einstein-Scrödinger equation combines mathematically quantum mechanics and general relativity.*

*Equation avoids the need for time dimension and it  describes universe as a wave function as needed.*

*Wheeler-DeWitt-equation describes the geometry of the universe and is  essential to understanding quantum gravity.*

*Quantum gravity describes events in which neighter gravitational nor quantum effects can be ignored like in the early stages of the universe.*

***At the scale close to Planck length quantum fluctuation play an important role.***

***Quantum fluctuation or vacuum fluctuation is the temporary random change in amount of energy in a point in space.***

***Quantum fluctuations are minute random fluctuations in the values of the fields which represent elemetary particles.***

***At the Planck scale the Standard Model, Quantum Field Theory and General Relativity must apply as the quantum effects of gravity dominate.***

***The First Premise of Quantum Gravity:***

***"The Quantum pressure work – Mass-energy Correspondence"***

***<u>"The Quantum Pressure Work of Graviton</u>***

***<u>Corresponds to</u>***

***<u>The Mass-energy of Graviton.</u>***

$$-Pqh \times \Delta Vbit = mbit \times c^2$$

***Pqh = Quantum holographic pressure***

***Vbit = Volume of gravitational bit (Graviton)***

***= Volume of Planck sphere***

$$= 4/3\pi\ (lp)^3 = \underline{1{,}7686 \times 10^{-104} m^3}$$

***mbit = Mass corresponding gravitation bit (Graviton)***

***c = Speed of light = 299792458m/s***

***lp (Planck lenght) = 1,61625518 x $10^{-35}$m***

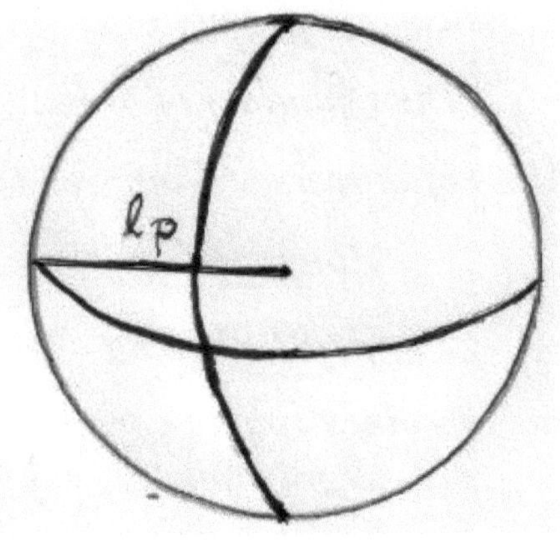

## ***Quantization of mass and volume***

***Mass corresponding to graviton:***

***The Quantum of Mass:***

***Mass of universe / Number of bits***

$$Mbit = Mu / N$$

$$= 2{,}991 \times 10^{-71} kg$$

***Volume of Planck sphere corresponding to graviton:***

***The Quantum of Volume***

$$Vbit = 4/3\pi \, (lp)^3$$

$$= 1{,}7686 \times 10^{-104} m^3$$

***Mass and volume corresponding to graviton.***

$$2{,}991 \times 10^{-71} kg \wedge 1{,}7686 \times 10^{-104} m^3$$

***Mass corresponding graviton makes volume corresponding graviton disappear from space.***

***Mass (m) pokes holes ($-\Delta V$) in the space!***

# The Universum

Volume $(V_U) \approx 4 \times 10^{80} m^3$

Energy Density $(\rho_U) \approx 9{,}9 \times 10^{-27} kg/m^3$

Mass $(M_U) \approx 3{,}96 \times 10^{54}$ kg

Mass-Energy $(E = mc^2) = 3{,}559 \times 10^{71} J$

Matter 4,9%

Dark Matter 26,8%

Dark Energy 68,3%

Gravitational radius $\left(r_s = \frac{2GM}{c^2}\right)$

$= 5{,}88 \times 10^{27} m$

Gravitational Horizon $\left(4\pi (r_s)^2\right)$

$= 4{,}346 \times 10^{56}$

Gravitational Information $\left(\frac{\left(\frac{2GM}{c^2}\right)^2}{l_p^2}\right)$

Planck length $(l_p) = 1{,}616255018 \times 10^{-35} m$

$N_{BIT} = 1{,}324 \times 10^{125}$

Planck Area $(A_p) = 4\pi (l_p)^2 = 3{,}2827 \times 10^{-69} m^2$

Planck sphere Volume $(V_p) = \frac{4}{3}\pi (l_p)^3$

$V_p = 1{,}7686 \times 10^{-104} m^3$

$m_{BIT} = \frac{M_U}{N_{BIT}} = 2{,}991 \times 10^{-71}$ kg

$V_{BIT} = V_p = 1{,}7686 \times 10^{-104} m^3$

# The Universum

Particle Horizon $= N_{BIT} \times \pi (l_P)^2$

$= 1{,}086 \times 10^{56}\, m^2$

Radius of Particle Horizon $= \sqrt{\dfrac{N_{BIT}\pi (l_P)^2}{4\pi}}$

$= 2{,}941 \times 10^{27}\, m$

Volume of Particle space

$$= \frac{4}{3}\pi \left(\sqrt{\frac{N_{BIT}\pi (l_P)^2}{4\pi}}\right)^3 = 1{,}06555 \times 10^{83}\, m^3$$

Energy:

Matter $= 1{,}744 \times 10^{70}\, J$

Dark Matter $= 9{,}538 \times 10^{70}\, J$

Dark Energy $= 2{,}421 \times 10^{71}\, J$

$$\rho_{DE} = \rho_{VAC} = \frac{\text{DARK ENERGY}}{V(\text{particle space})}$$

$$= 2{,}264 \times 10^{-12}\, J/m^3$$

## cosmological constant

$$\Lambda = \frac{8\pi G}{c^4}\rho = 4{,}701 \times 10^{-55}\, 1/m^2$$

## ***The Mass of The Universe***

***Volume of the universe*** $Vu = 4x10^{80}m^3$

***Energy density of the universe*** $\rho E = 9{,}9 \times 10^{-27} kg/m^3$

***Mass of universe*** $Mu = Vu \times \rho E = \underline{3{,}96 \times 10^{54} kg}$

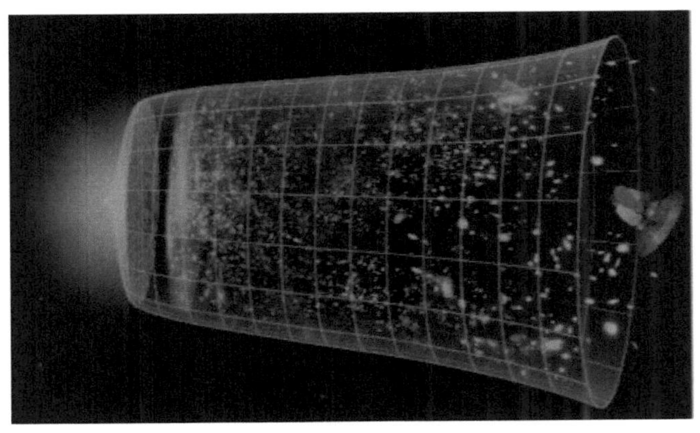

***Picture: NASA***

The Mass-radius

$$V = \frac{m}{m_{BIT}} \times \frac{4}{3} \pi l_p^3$$

$$r = \sqrt[3]{\frac{m}{m_{BIT}} l_p^3}$$

Black = Mass radius

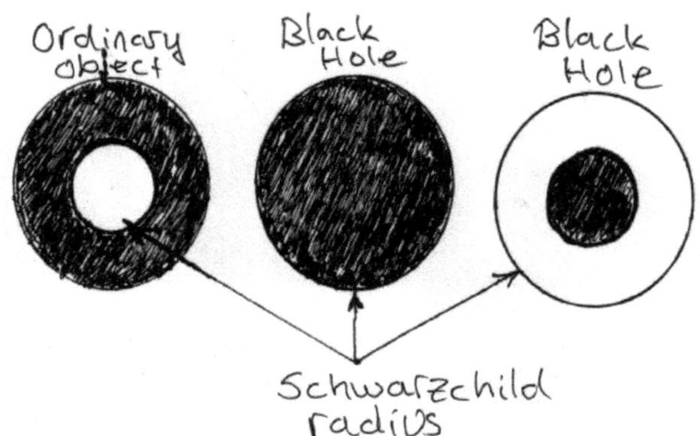

## *Schwarzschild cosmology*

***Schwarzschild cosmology also called black hole cosmological model is a cosmological model in which the observable universe is the interior of a black hole.***

## *The Gravitational radius*

***The Schwarzschild radius or the gravitational radius is a physical parameter that corresponds to the radius defining the event horizon of a Schwarzschild black hole. Every mass has a Schwarzschild radius***

***The Schwarzschild radius rs is given as***

$$rs = (2Gm)/c^2,$$

***G = Gravitational constant = 6,67348 x*** $10^{-11} m^3/kgs^2$

***m = The object mass***

***c = Speed of light***

***The Ordinary Object´s Mass-radius***

***> The Schwarzschild radius.***

***The Black Hole Mass-radius***

***≤ The Schwarzschild radius.***

Smallest Black Hole

$$r_S = r_M$$

$$\frac{2Gm}{c^2} = \sqrt[3]{\frac{m}{m_{BIT}} l_P^3}$$

$$m \approx 3 \times 10^{23}\, kg$$

## ***The gravitational quantum information is coded on gravitational horizon.***

***The gravitational radius defines gravitational horizon.***

***Area of gravitational horizon:***

$$4\pi \left((2GM)/c^2\right)^2$$

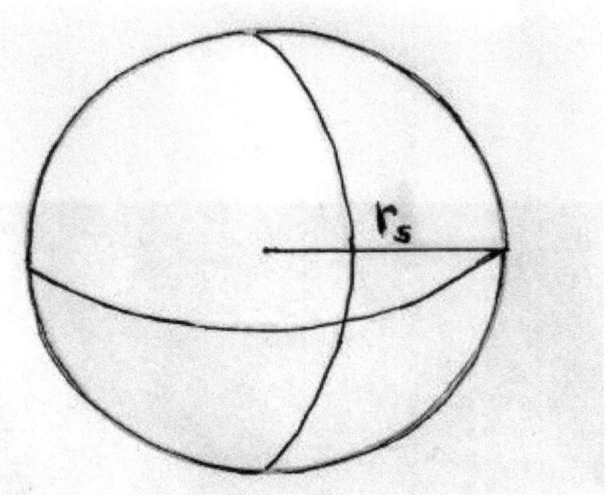


***Gravitational radius (Rs) of the universe:***

$$Rs = (2GMu)/c^2$$

$$= 5{,}881 \times 10^{27} m$$

***Gravitational horizon area (Agh) of the universe***

$$Agh = 4\pi\,((2GMu)/c^2)^2$$

$$= 4{,}346 \times 10^{56} m^2$$

## *The Bekenstein Bound*

***The Bekenstein bound is an upper limit of entropy, that can be contained whitin finite ragion of space which has finite amount of energy – or conversely, the maximal amount of information required to perfectly describe the physical system to quantum level.***

***It implies that the information necessary to perfectly describe that system, must be finite if the region of space and the energy are finite***

***Bekenstein-Hawking entropy***

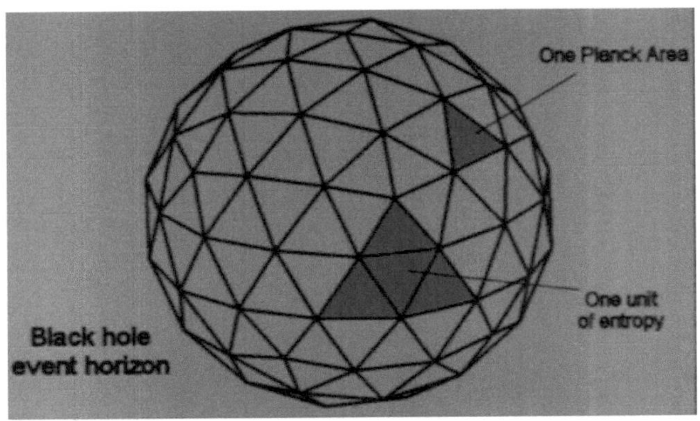


***Photo: Jacob D. Bekenstein***

## ***One Planck volume area = 1 bit of information***

***One Planck area or great circle area of Planck volume***

***Planck area =*** $\pi lp^2$

$= 8{,}207 \times 10^{-70} m^2$

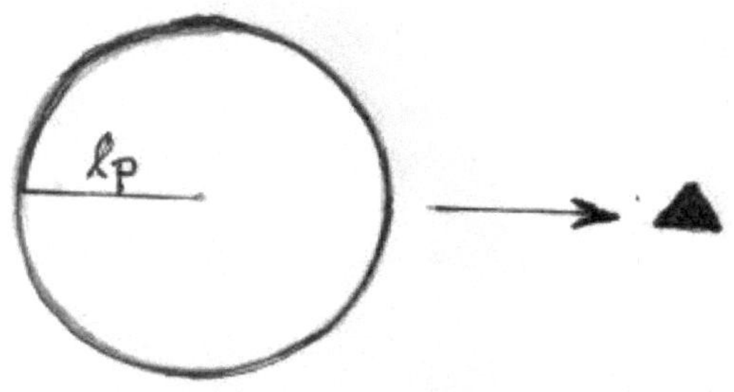

***4 x Planck areas = One Planck volume area***

***= one unit entropy = 1 bit of information.***

$\underline{4 \times \pi lp^2 = 4\pi lp^2}$

$= \underline{3{,}284 \times 10^{-69} m^2}$

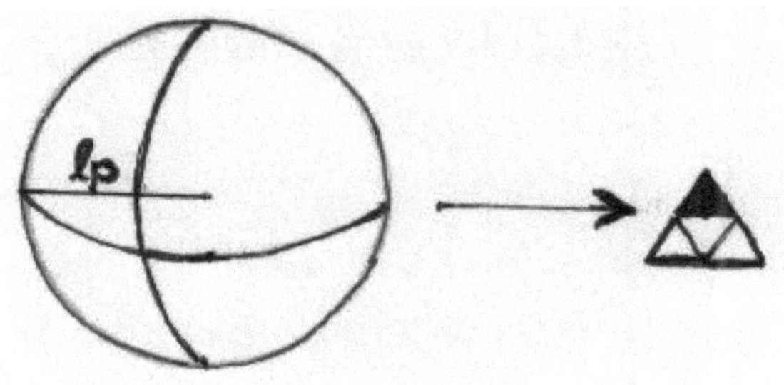


***The gravitational quantum information is coded on gravitational horizon.***

***Number of gravitational information bits on gravitational horizon:***

***Area of gravitational horizon / Area of one Planck volume***

$$N = (4\pi\ ((2GMu)/c^2)^2)\ /\ (4\pi\ lp^2)$$

$$= 1{,}324 \times 10^{125} \text{ bits of information}$$

$$= \underline{1{,}324 \times 10^{125}\ \text{gravitons}}$$

***The Graviton***

***<u>The Quantum Bit of Gravity</u>:***

***Mass of universe / Number of bits***

***mbit = Mu / N***

***corressponds to <u>2,991 x10⁻⁷¹kg</u>***

***In theories of quantum gravity, the graviton is the hypothetical quantum of gravity, an elementary partice that mediates the force of gravitational interaction.***

***<u>Graviton is a massless tensor boson corressponding to certain area, mass and volume.</u>***

***Graviton is geometrically Planck sphere, a little bubble with Planck length radius that decreases the volume of space which causes distorsion in space.***

$$Vbit = 4/3\pi\ (lp)^3 = 1{,}7686x\ 10^{-104}m^3$$

## *Gravitational particle horizon*

***Gravitational particle horizon represents the largest distance where gravitons can mediate the force of gravitational attraction. At larger distances graviton bubbles will burst and lose their volume. At this maximal distance gravitons form a quantum foam horizon of quantum bubbles where all bubbles are side by side. Every graviton has radius of Planck length so every quantum foam bubble takes area of $\pi\ lp^2$ from the horizon.***

*Area of gravitational particle horizon*

$$Agh = N \times \pi\ lp^2$$

$$= 1{,}0866 \times 10^{56} m^2$$

***The radius of gravitational particle horizon***

$$Rgh = \sqrt{(N \times \pi\ lp^2 / 4\pi)}$$

$$= \underline{2{,}941 \times 10^{27} m}$$

## *Gravitational Potential Energy*

***Gravitational potential energy is the potential energy a massive object has in relation to another massive object due to gravity.***

***For two pairwise interacting point particles (M and m), the gravitational potential energy U is given by***

$$U = -(G \times M \times m)/r$$

***In zero-energy universe sum of mass-energy and sum of gravitational potential equals zero.***

$$\sum mc^2 + \sum -(G\,Mm)/r = 0$$

$$Mu \times c^2 = (G \times Mu \times N \times Mbit) / Rgpe$$

***<u>Radius of gravitational potential energy</u>***

$$Rgpe = (G \times N \times Mbit) / c^2$$

$$= \underline{2{,}941 \times 10^{27} m}$$

***Radius of gravitational potential energy is distance between gravitational singularity and gravitational particle horizon***

## *The graviton horizon radius*

$$-Pqh \times \Delta Vbit = mbit \times c^2$$

$$mbit = -Pqh \times \Delta Vbit / c^2$$

$$Rgh$$

$$= \sqrt{(N \times \pi\, lp^2 / 4\, \pi)} = (-G \times N \times Mbit) / c^2$$

$$= (-G \times -Pqh \times N \times \Delta vbit) / c^4$$

$$= (-G \times -Pqh \times N \times 4/3\pi\, (lp)^3\,) / c^4$$

$$= \underline{2{,}941 \times 10^{27} m}$$

## *The Second Premise of Quantum Gravity*

## *"The Radius Equivalence"*

***The radius of particle horizon***

$$=\sqrt{(N \times \pi lp^2 / 4 \pi)}$$

***Equals***

***The radius of gravitational potential energy***

$$= (-G \times N \times Mbit) / c^2$$

***Equals***

***The radius of graviton horizon***

$$= (-G \times -Pqh \times N \times 4/3\pi (lp)^3) / c^4$$

***Equals***

$$= 2{,}941 \times 10^{27} m$$

***The Volume of Gravitons***

***"The Lost Volume of Space"***

$$= N \times 4/3\pi (lp)^3$$

$$= 2{,}342 \times 10^{21} m^3$$

***(If there would not be Quintessence!)***

## *Quantum pressure*

## ***Quantum holographic pressure:***

***Pqh** = $(c^4 \times \sqrt{(N \times \pi\, lp^2 / 4\pi)}) / (N \times -4/3\pi\, (lp)^3 \times -G)$*

***Eguals***

$$\frac{(\text{Speed of light})^4 \times (\text{Graviton horizon radius})}{(\text{Gravitational constant}) \times (\text{The Volume of Gravitons})}$$

= ***$1{,}52 \times 10^{50}$ Pa***

## *The Third Premise*

## *"The Equation of Quantum Gravity":*

***Quantum holographic pressure =***

$$\frac{(\textit{Speed of light})^4 \times (\textit{Graviton horizon radius})}{(\textit{Gravitational constant}) \times (\textit{The Volume of Gravitons})}$$

***Speed of light → Quantum***

***Graviton horizon radius → Particle / 2D - Area***

***Gravitational constant → Gravity***

***Quanton volume → Particle / 3D Volume***

***Positive quantum pressure causes loss in the space volume.***

***Matter and Dark Matter have positive quantum pressure, gravitational attraction and negative gravitational potential energy.***

## *Conclusion:*

***GRAVITONS ARE MASSLESS TENSOR BOSONS THAT CAUSE GRAVITATIONAL ATTRACTION AND DISTORSION IN SPACE-GEOMETRY.***

***Can we add the Graviton to The Standard Model of Particle Physics?***

## *Holographic principle*

***"Three-dimensional world of ordinary experience – the universe filled with galaxies stars, planets, houses, boulders and people – is a hologram, an image of reality coded on a distant two-dimentional surface"***

***– Leonard Susskind***

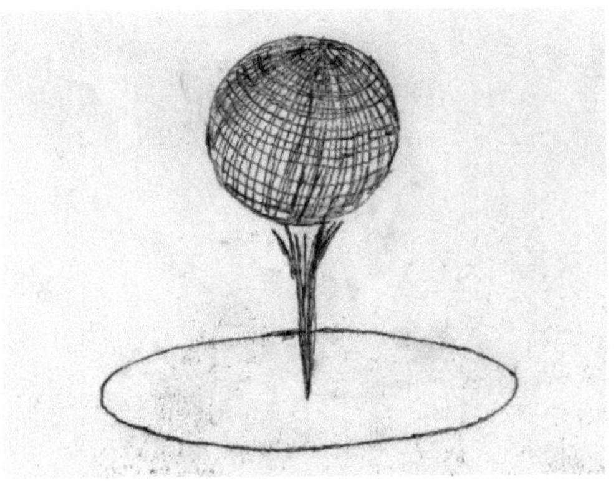

***Holographic principle is a property of quantum gravity that states that the description of a volume on space is coded on a lower-dimensional boundry to the region – such as a light-like boundry like gravitational horizon.***

*Riemann sphere*

***"The space and time that we percieve in large scale are our blurred and approximate image of the gravitational field."***

***- Carlo Rovelli***

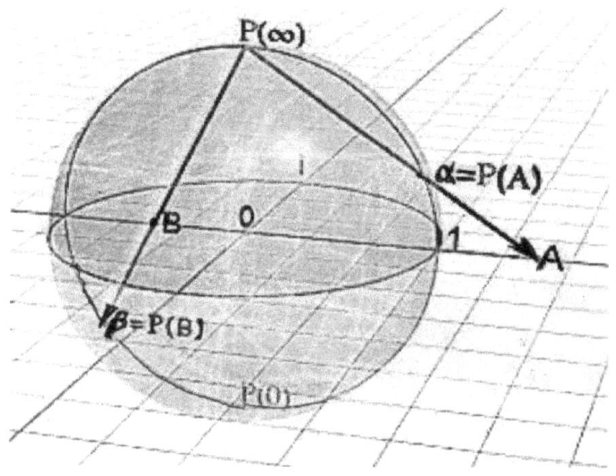


***Picture: Jean-Christophe Benoist***

***Riemann sphere is origocentered sphere on complex plane.***

***Every point in Riemann sphere surface (for example α and β) is projected to complex plane (for example A and B).***

## ***Riemann sphere and holographic gravitational field***

***"Information is energy in formation"***

***- Daniel Siegel***

***2 D gravitational particle horizon is considered as Riemann sphere.***

***3D-coordinates (x, y, z) of gravitons are coded to 2D gravitational horizon as quantum information.***

***Quantum information code i.e. spatial euclidian 3D-coordinates (x, y, z) of gravitons is projected to 2D complex plane as gravitational field.***

***The Wheeler-DeWitt-equation could describes the 3-dimensional geometry of the universe.***

***This is example of holographic principle.***

***At the fundamental level reality is not made of separate particles but interconnected whole like an hologram.***

***Mass makes Space-Grid distorded.***

***Mass (m) pokes bubbles ($-\Delta V$) in the space***

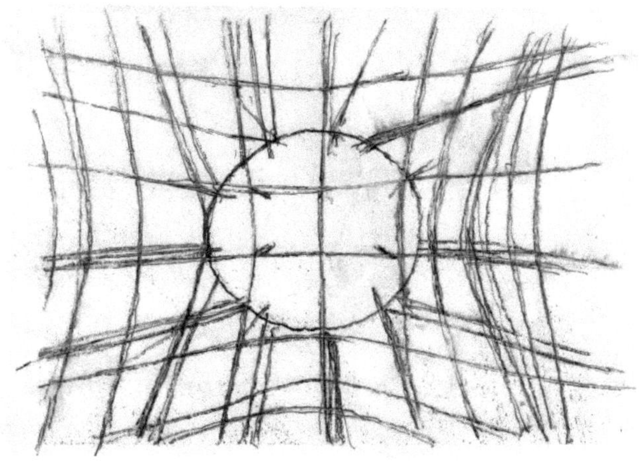

## ***Quantization of Space***

***Volume of particle Space Grid***

$$Vsg = 4/3\pi\ (Rgph)^3$$

$$= 1{,}06555 \times 10^{83} m^3$$

***Voxel of Space-Grid***

$$Vvox = lp^3$$

$$= 4{,}222 \times 10^{-105} m^3$$

***Voxels in Space-Grid***

$$Nvox = Vpu / Vvox$$

$$= 4/3\pi\, (Rgpe)^3 / lp^3$$

$$= 2{,}524 \times 10^{187} \text{ voxels}$$

***How many voxels correspond to 1 graviton?***

$$4/3\pi\, (lp)^3 / lp^3$$

$$= 4{,}189 \text{ voxels}$$

***Background space is a grid of voxels like a huge Rubik´s cube with euclidean geometry without distorsion.***

***Gravitational background space is three-dimensional space where particles can be located by three spatial coordinates (x, y, z).***

***Mass makes space in Space-Grid distorded.***

## *The Quintessence*

***The Quintessense is Dark Energy, a form of energy that causes the expansion of the universe.***
***It increases the volume of space inside the space-grid.***
***Quintessence has been proposed to be the fifth fundamental force.***
***Quantum of Quintessence is Quinton.***
***Quintons have Quantum vacuum and repulsive force.***
***The mass of Quintessence is negative and it has positive gravitational potential energy.***
***Gravitational potetial energy:***

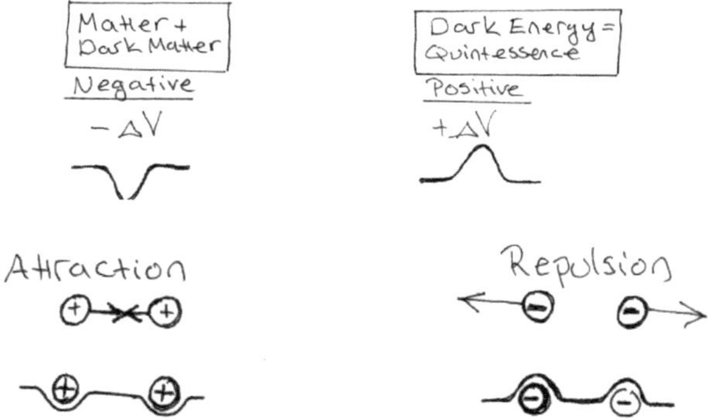

## *Mass and negative mass*

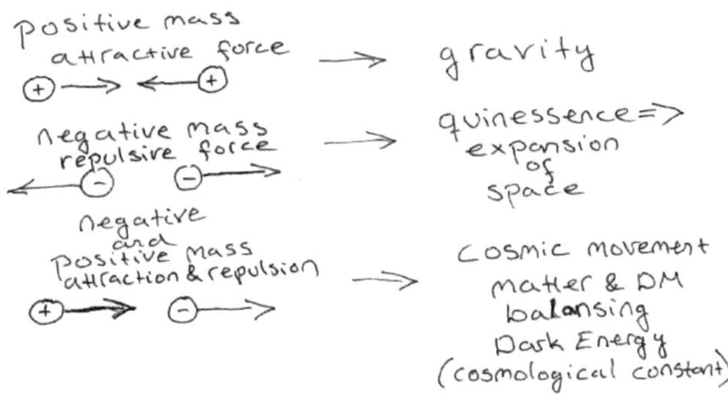


## *Negative Mass*

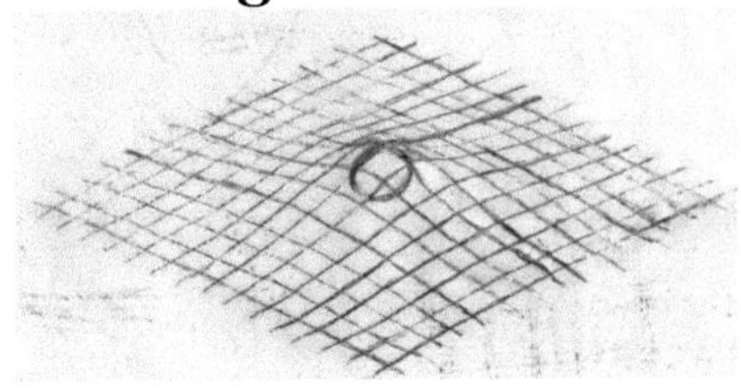

## *Positive Mass*

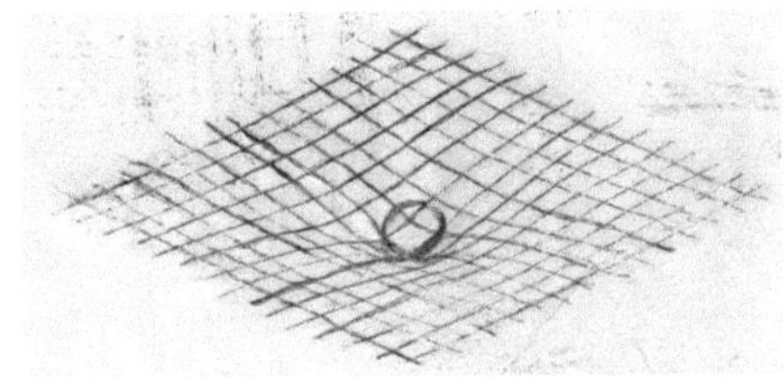

## *Mass-energy of the Universe*

$$Mu = 3{,}96 \times 10^{54} kg \rightarrow 3{,}532 \times 10^{71}\ J$$

→ *4,9 Matter*
→ *26,8% Dark matter*
→ *68,3% Dark energy*

## *Mass-energy of Dark energy*

$$68{,}3\ \% \times 3{,}532 \times 10^{71}\ J$$

$$= 2{,}412 \times 10^{71} J$$

## *Vacuum energy density of the universe*

***Mass-energy of Dark energy / Volume of particle space grid***

## *Dark Energy Density:*

$$\rho\ vac\ = 2{,}264 \times 10^{-12} J/m^3$$

## ***The Cosmological Constant***

***Einstein´s cosmological constant is proportional to Dark Energy Density***

$$\Lambda = (8\,\pi\,G\ \ x\,\rho) / c^4$$

$$\underline{= 4{,}701 \ x \ 10^{-55}\ 1/s^2}$$

## ***Vacuum Energy Density equals volume expansion***

$$\rho \; vac = 2{,}264 \times 10^{-12} J/m^3$$

*1 $m^3$ corresponds to 2,519 x $10^{-29}$kg*

*1 $m^3$ ^ 1,490 x $10^{-62}m^3$ (Expansion)*

*1 $m^3$ ^ 8,422 x $10^{41}$ Quintons*

***In whole particle universe volume expansion is***

*V exp*

$$1{,}49 \times 10^{-62} m^3/m^3 \times 1{,}0655 \times 10^{83} m^3$$

$$\underline{1{,}59 \times 10^{21} m^3}$$

## ***Time and Frequency***

***Planck time is a natural unit of time, equivalent to the time it takes light to traverse one Planck lenght: its is the smallest duration of time that has physical meanining. Planck time is known as "Quantum of time".***

***Planck time = 5,39 x $10^{-44}$s***

***An event that occurs once every Planck time has Planck frequency. It is said to be upper bound for frequency of cosmic rays.***

***Planck frequency = 1,855 x $10^{43}$Hz***

***The wavelength of the emitted radiation is inversely proportional to its frequency***

$$\lambda = c / f$$

***Planck wavelength = 1,6163 x $10^{-35}$m***

***= Planck length***

***1 Planck lenght to every direction in space increases space volume by Planck volume:***

***$4/3\pi\ (lp)^3 = 1{,}7686x\ 10^{-104}m^3$***

***<u>Space volume increased in 1 second:</u>***

***Uve = Universe volume expansion x Planck frequency***

***<u>$= 2{,}944 \times 10^{64} m^3/s$</u>***

***<u>$6{,}39 \times 10^{13}$ x Speed of light!!!</u>***

***<u>Space Expansion and Age of the Universe</u>***

***"Light beams became alive, and become not only alive, but self-awere, and acquired the ability to wonder. The wonder is not wheter this genesis took six days or fourteen billion years or even eternity"***

***- Gerald Schroeder***

***<u>Space volume increased in 1 year:</u>***

***Vs1y = Uve x 3156000s / year***

***<u>$= 9{,}284 \times 10^{71} m^3/y$</u>***

## ***Age of the Universe***

***(U-age)***

***= Universe volume / Vs1y***

***= $4x10^{80}m^3$ / $1,547 \times 10^{70}m^3/y$***

***430 x $10^6$ years***

***Measurements on direct observations indicate that The Age of the Universe is 13,8 billion years.***

***According to these calculation Universe has been expanding faster than light from the very Beginning.***

## ***In The Beginning there was Energy***

## ***and Symmetry...***

***"It is important to realize that in physics today, we have no knowledge what energy is"***

***– Richard Feynman***

***Smallest possible Universe:***

$$4/3\pi\,(lp)^3 = 1{,}7686 \times 10^{-104} m^3$$

***Vacuum Energy density***

$$\rho\ vac = 2{,}264 \times 10^{-12} J/m^3$$

# ***THERE WAS ENERGY AT THE BEGINNING!***

***E0***

***Equals***

***$4/3\pi\ (lp)^3 \times \rho\ vac$***

***$= 4{,}004 \times 10^{-116} J$***

***V0***

***Equals***

***$4/3\pi\ (lp)^3 = 1{,}7686 \times 10^{-104} m^3$***

***m0***

***$4/3\pi\ (lp)^3$ corresponds to mbit***

***$= 2{,}991 \times 10^{-71} kg$***

## ***Quantum fluctuation I:***

***Energy 1***

***Equals***

***$2{,}991 \times 10^{-71} kg \times c^2$***

***$= 2{,}688 \times 10^{-54} J$***

***Volume 1***

***Equals***

***$E1 / \rho\ vac$***

***$= 1{,}187 \times 10^{-42} m^3$***

***Mass 1***

***Equals***

***$V1/(4/3\pi\ lp^3) \times mbit$***

***$= 2{,}008 \times 10^{-9} kg$***

# ***Quantum fluctuation II***

***Energy 2***

***Equals***

***$m1 \times c^2$***

***$= 1{,}8047 \times 10^8 J$***

***Volume 2***

***$E2 / \rho\ vac$***

***$=7{,}971 \times 10^{19} m^3$***

***Mass 2***

***$V2 / (4/3\pi\ lp^3) \times mbit$***

***$= 1{,}348 \times 10^{53} kg$***

***SYMMETRY***

***E vac***

***$\Delta V$***

***$mc^2$***

***SYMMETRY BREAK:***

***Mass 2 < Mass of the Universe***

***$1{,}348 \times 10^{53}kg < 3{,}96 \times 10^{54}kg$***

***Total Volume of Mass-energy (Vtot)***

***$= 3{,}96 \times 10^{54}kg / mbit \times 4/3\pi\ lp^3$***

***$= 2{,}342 \times {}^{21}m^3$***

***Volume lost with Matter and Dark Matter***

***$= (4{,}9\% + 26{,}8) \times Vtot$***

***$= 7{,}42 \times 10^{20}m^3$***

***Volume of Dark Matter***

***$68{,}3\% \times V\ tot$***

***$= 1{,}59 \times 10^{21}m^3$***

***<u>THE VOLUME OF DARK MATTER EQUALS THE PARTICLE UNIVERSE EXPANSION:</u>***

***68,3 % OF TOTAL VOLUME OF MASS-ENERGY***

***EQUALS***

***VOLUME EXPANSION:***

$$1{,}49 \times 10^{-62} m^3/m^3 \times 1{,}06555 \times 10^{83} m^3$$
$$\underline{1{,}59 \times 10^{21} m^3}$$

***CONCLUSION:***

***<u>QUANTUM FLUCTUATIONS OF DARK ENERGY CAUSE EXPANSION OF THE UNIVERSE!</u>***

## ***Singularity***

***In the beginning there was limited extremely dence mass called singularity.***

***After symmetry break of gravitational singularity gravitons i.e. gravitation separated from other  Fundamental Forces except from Dark Energy or Quintessence.***

***Quintessence is considered as Fifth Fundamental Force.***

***Maybe Quintessence or Dark Energy should be in the Standard Model of Particle physics.***

## ***Cosmic inflation***

***Cosmic inflation is an exponential expansion of space in the Beginning of the Universe.***

***Volume 0 = $4/3\pi\ (lp)^3 = 1{,}7686 \times 10^{-104} m^3$***

***Volume 1 = $1{,}187 \times 10^{-42} m^3$***

***Volume 2 = $7{,}971 \times 10^{19} m^3$***

***SYMMETRY BREAK!***

***Volume 3 = $1{,}59 \times 10^{21} m^3$***

***Volume 4 = Volume 3 x 2***

***Volume 5 = Volume 3 x 3***

***etc.***

***Expansion At the Beginning:***

***$V0 \rightarrow V1 \equiv 6{,}7 \times 10^{61}$ x***

***$V1 \rightarrow V2 \equiv 6{,}7 \times 10^{61}$ x***

***$V2 \rightarrow V3 \equiv 0{,}199$ x***

***$V3 \rightarrow V4 \equiv 2$ x***

***$V4 \rightarrow V5 \equiv 1{,}5$ x***

***etc***

## ***Black Hole Singularity***

***Smallest Black Hole ≈ 3 x $10^{23}$kg***

***Smallest known Black Hole is 3x the mass of the Sun.***

***≈ 6 x $10^{30}$kg***

***Example:***

***Black Hole Mass: $10^{32}$kg***

The Mass-radius

$$V = \frac{m}{m_{BIT}} \times \frac{4}{3} \pi l_p^3$$

$$r = \sqrt[3]{\frac{m}{m_{BIT}} l_p^3}$$

***r Mass = 0,2416m***

***r Schwarzschild = (2Gm)/c² = 148513m***

***The Black Hole Mass-radius***

***≤ The Schwarzschild radius.***

## *Volume of Black Hole Singularity*

***Black Hole Mass: $10^{32}$kg***

***<u>V singularity:</u>***

***$10^{32}$kg / mbit x $4/3\pi\ (lp)^3$***

***<u>$=0{,}0591m^3$</u>***

***<u>Total Black Hole volume:</u>***

***$4/3\pi$ (r Schwarzschild)$^3$***

***<u>$= 1{,}37 \times 10^{16}m^3$</u>***

***Space is distorded inside Black hole. There is singularity size empty volume in space-grid of the Black Hole. The rest of the Black Hole volume is Dark Energy or Quintessence. Dark Energy creates negative pressure inside the Black Hole. Negative pressure creates Graviton radiation by filling singularity.***

## ***Hawking radiation***

***Temperature of Black Hole radiation:***

$$TH = \frac{\hbar c^3}{8 \pi G M kb}$$

***ħ = reduced Planck constant***
***= 1,054571817 x 10⁻³⁴Js***

$\hbar$ = reduced Planck constant
$= 1{,}054571817 \times 10^{-34} Js$
$kb$ = Boltzmann constant
$= 1{,}380649 \times 10^{-23} m^2kg/s^2K$

***For excample***

***Black Hole Mass:*** $10^{32}kg$

$TH = 1{,}227 \times 10^{-9}K$

***Hawking Radiation Energy***

$Q = m \times c \times \Delta T$

$m$ = Mass (kg)

$TH = \Delta T$ = Temperature (K)

$C = 4{,}184 kJ/kgK$

***Black Hole Mass:*** $10^{32}kg$

$TH = 1{,}227 \times 10^{-9}K$

$Q = 5{,}134 \times 10^{26}J$

$Q = \Delta m\, c^2$

$\Delta m = Q / c^2$

$\Delta m = 5{,}713 \times 10^{9}kg$

$N$ graviton $= \Delta m / mbit$

$= 1{,}91 \times 10^{80}$ gravitons

***Black Hole loses it´s mass as graviton radiation.***

***CONCLUSION:***

***<u>HAWKING RADIATION IS GRAVITON RADIATION</u>***

## *Gravitational Waves*

***Gravitational Waves are coherent waves of Gravitons.***

***They are invisible and incredibly fast***

***ripples in space. (v = c)***

***"The detection of a single graviton may in fact be ruled out in real universe"***

***- Freeman John Dyson***

## *The Mass of Neutrino*

***The mass of electron neutrino:***

***1,25 x $10^{-37}$kg***

***Mass-volume:***

$$= \frac{m\ neutrino \times 4/3\pi\ (lp)^3}{mbit}$$

***= 7,391 x $10^{-71}$m³***

***Schwarzchild volume:***

$$4/3\ \pi\ ((2 \times m\ neutrino \times G)\ /\ c^2)^3$$

***= 2,680 x $10^{-191}$m³***

***The Electron neutrinos´s Mass-radius***

***> The Schwarzschild radius.***

***Neutrino makes hole in space-grid***

$$= \frac{7{,}391 \times 10^{-71} m^3}{lp^3}$$

***= 1,75 x $10^{34}$ voxels***

***"One who claims to be skeptic of one set of beliefs is actually a true believer in another set of beliefs."***

***– Phillip E. Johnson.***

*Thanks to My Loving Wife Maria who allways supported me.*

*Thanks to our three Wonderful Children who inspired me to write this book.*

*Thanks to my Parents and Relatives who raised me and took care ofme.*

*Thanks to my Excellent Teachers who educated me.*

*And special thanks to my Colleagues who helped me through hard times and made this thought process possible.*

# *Notes*

# *Notes*

# *Notes*

***Notes***

## *Notes*